2019年

全国水利发展统计公报

2019 Statistic Bulletin on China Water Activities

中华人民共和国水利部　编

Ministry of Water Resources, People's Republic of China

中国水利水电出版社
www.waterpub.com.cn
·北京·

图书在版编目（CIP）数据

2019年全国水利发展统计公报 = 2019 Statistic Bulletin on China Water Activities / 中华人民共和国水利部编. -- 北京 : 中国水利水电出版社, 2020.11
ISBN 978-7-5170-9092-2

Ⅰ. ①2… Ⅱ. ①中… Ⅲ. ①水利建设－经济发展－中国－2019 Ⅳ. ①F426.9

中国版本图书馆CIP数据核字(2020)第213316号

书 名	**2019 年全国水利发展统计公报** **2019 Statistic Bulletin on China Water Activities** 2019 NIAN QUANGUO SHUILI FAZHAN TONGJI GONGBAO
作 者	中华人民共和国水利部 编 Ministry of Water Resources, People's Republic of China
出版发行	中国水利水电出版社 （北京市海淀区玉渊潭南路 1 号 D 座 100038） 网址：www. waterpub. com. cn E-mail：sales@ waterpub. com. cn 电话：(010) 68367658（营销中心）
经 售	北京科水图书销售中心（零售） 电话：(010) 88383994、63202643、68545874 全国各地新华书店和相关出版物销售网点
排 版	中国水利水电出版社微机排版中心
印 刷	北京印匠彩色印刷有限公司
规 格	210mm×297mm 16 开本 3.75 印张 52 千字
版 次	2020 年 11 月第 1 版 2020 年 11 月第 1 次印刷
印 数	0001—1000 册
定 价	**39.00** 元

目　录

Contents

2019 年是中华人民共和国成立 70 周年，是决胜全面建成小康社会的关键一年，也是践行水利改革发展总基调的开局之年。党中央、国务院高度重视水利工作。一年来，在习近平新时代中国特色社会主义思想指引下，各级水利部门认真贯彻落实党中央、国务院决策部署，坚持以政治建设为统领，坚持水利改革发展总基调，坚持目标导向、问题导向、结果导向，真抓实干、攻坚克难，推动各项水利工作取得明显成效。

1 水利固定资产投资

2019年，水利建设完成投资6711.7亿元，较上年增加109.1亿元，增加1.7%。其中：建筑工程完成投资4987.9亿元，较上年增加2.3%；安装工程完成投资243.1亿元，较上年减少13.5%；机电设备及各类工器具购置完成投资221.1亿元，较上年增加3.1%；其他完成投资（包括移民征地补偿等）1259.7亿元，较上年增加2.4%。

	2012年/亿元	2013年/亿元	2014年/亿元	2015年/亿元	2016年/亿元	2017年/亿元	2018年/亿元	2019年/亿元	2019年比上年增加比例/%
全年完成	3964.2	3757.6	4083.1	5452.2	6099.6	7132.4	6602.6	6711.7	1.7
建筑工程	2736.5	2782.8	3086.4	4150.8	4422.0	5069.7	4877.2	4987.9	2.3
安装工程	237.8	173.6	185.0	228.8	254.5	265.8	280.9	243.1	-13.5
设备及各类工器具购置	178.1	161.1	206.1	198.7	172.8	211.7	214.4	221.1	3.1
其他（包括移民征地补偿等）	811.8	640.2	605.6	873.9	1250.3	1585.2	1230.1	1259.7	2.4

在全年完成投资中，防洪工程建设完成投资2289.8亿元，较上年增加5.3%；水资源工程建设完成投资2448.3亿元，较上年减少4.0%；水土保持及生态工程完成投资913.4亿元，较上年增加

23.2%；水电、机构能力建设等专项工程完成投资1060.2亿元，较上年减少6.7%。

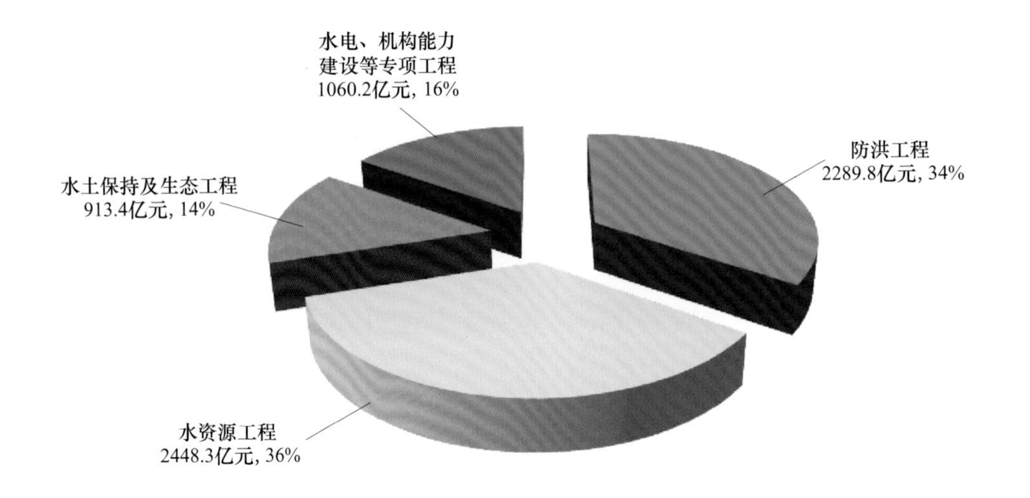

七大江河流域完成投资5262.1亿元，东南诸河、西北诸河以及西南诸河等其他流域完成投资1449.6亿元；东部、中部、西部、东北地区完成投资分别为2593.4亿元、1653.2亿元、2245.2亿元和219.9亿元。

在全年完成投资中，中央项目完成投资66.4亿元，地方项目完成投资6645.3亿元。大中型项目完成投资1066.8亿元，小型及其他项目完成投资5644.9亿元。各类新建工程完成投资5202.8亿元，扩建、改建等项目完成投资1508.9亿元。

全年水利建设新增固定资产3473.8亿元。截至2019年年底，在建项目累计完成投资17649.4亿元，投资完成率为62.7%；累计新增固定资产9987.6亿元，固定资产形成率为56.6%，较上年增加0.1个百分点。

当年在建的水利建设项目28742个，在建项目投资总规模28166.9亿元，较上年增加2.4%。其中：有中央投资的水利建设项目17202个，较上年增加1.6%；在建投资规模14058.3亿元，较上年减少1.0%。新开工项目20779个，较上年增加5.0%，新增投资规模6868.0亿元，比上年增加13.3%。全年水利建设完成土方、石方和混凝土方分别为29.9亿立方米、3.9亿立方米、1.9亿立方米。截至2019年年底，在建项目计划实物工程量完成率分别为：土方98.7%、石方94.3%、混凝土方95.5%。

水利固定资产完成投资

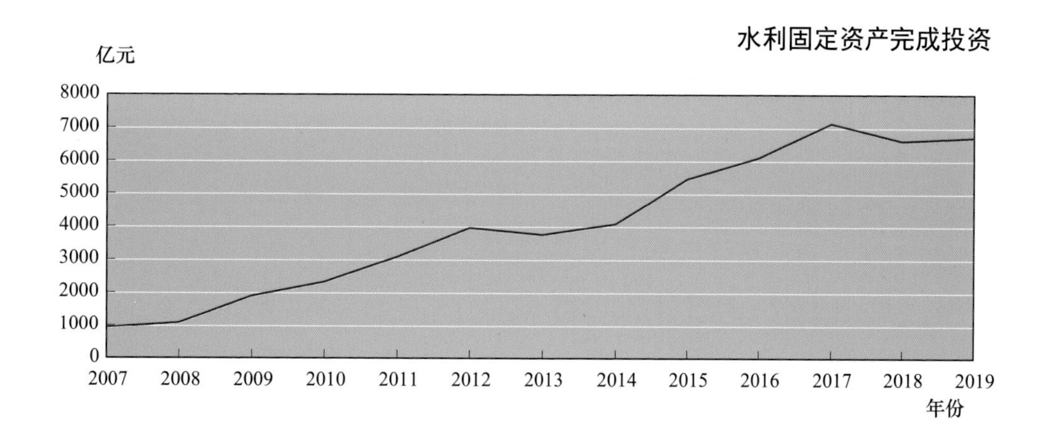

2 重点水利建设

江河湖泊治理。2019年，在建江河治理工程5414处，其中：堤防建设601处、大江大河及重要支流治理739处、中小河流治理3331处、行蓄洪区安全建设及其他项目743处。截至2019年年底，在建项目累计完成投资3882.9亿元，项目投资完成率为67.2%。长江中下游河势控制和河道整治工程有序实施；黄河下游防洪工程基本完工；进一步治淮38项工程已开工32项，其中8项建成并发挥效益；洞庭湖、鄱阳湖治理深入推进；太湖流域水环境综合治理21项工程已开工19项，其中10项已建成并发挥效益。

水库及枢纽工程建设。2019 年，在建水库及枢纽工程 1360 座，截至 2019 年年底，在建项目累计完成投资 3197.5 亿元，项目投资完成率为 60.3%。贵州凤山水库、新疆玉龙喀什水利枢纽、江西花桥水利枢纽、四川龙塘水库、海南迈湾水利枢纽、四川江家口水库等大型水库及枢纽和西音水库、秀水水库等 5 座中型水库开工。大藤峡水利枢纽、新疆大石峡水利枢纽、湖北碾盘山水利枢纽、四川土溪口水库、四川李家岩水库等工程实现年度导截流目标；河南出山店水库、新疆阿尔塔什水利枢纽、西藏拉洛水利枢纽、河北双峰寺水库、山东庄里水库、湖南莽山水库、黑龙江奋斗水库、贵州马岭水库等工程下闸蓄水。

水资源配置工程建设。2019 年，水资源配置工程在建投资规模为 7875.0 亿元，累计完成投资 4530.2 亿元，项目投资完成率为 57.5%。广东珠江三角洲水资源配置等工程开工建设；湖北鄂北地区水资源配置工程封江口水库上游段、江西廖坊水利枢纽灌区二期工程实现通水，引江济淮、云南滇中引水、内蒙古引绰济辽、陕西引汉济渭、贵州夹岩水利枢纽及黔西北调水等工程加快实施。

农村水利建设。2019 年，农村饮水安全巩固提升工程完成投资 619.6 亿元，其中中央补助资金 121.7 亿元。当年安排中央投资用于大中型灌排工程建设与灌区节水改造 86 亿元，重点中型灌区节水改造 63.5 亿元，高效节水灌溉等农田水利建设 79.6 亿元。全年新增耕地灌溉面积 780 千公顷，新增节水灌溉面积 1076 千公顷，新增高效节水灌溉面积756千公顷。截至2019年年底，农村自来水普及率达到82%，农村集中供水率达到87%。

农村水电建设。2019年，农村水电建设完成投资71.0亿元，新增水电站90座，装机容量107.2万千瓦，其中：新投产装机容量56.4万千瓦，技改净增发电设备容量50.8万千瓦。

水土流失治理。2019年，水土保持及生态工程在建投资规模为1090.8亿元，累计完成投资913.4亿元。全国新增水土流失综合治理面积6.7万平方公里，其中国家水土保持重点工程新增水土流失治理面积1.1万平方公里。对600座黄土高原淤地坝进行了除险加固。

行业能力建设。2019 年，水利行业能力建设完成投资 114.9 亿元，其中：防汛通信设施投资 8.5 亿元，水文建设投资 18.8 亿元，科研教育设施投资 1.6 亿元，其他投资 86.0 亿元。

3 主要水利工程设施

堤防和水闸。截至2019年年底，全国已建成5级及以上江河堤防32.0万公里[1]，累计达标堤防22.7万公里，达标率为71.0%；其中1级、2级达标堤防长度为3.5万公里，达标率为81.7%。全国已建成江河堤防保护人口6.4亿人，保护耕地4.2万千公顷。全国已建成流量5立方米每秒及以上的水闸103575座，其中大型水闸892座。按水闸类型分，分洪闸8293座，排（退）水闸18449座，挡潮闸5172座，引水闸13830座，节制闸57831座。

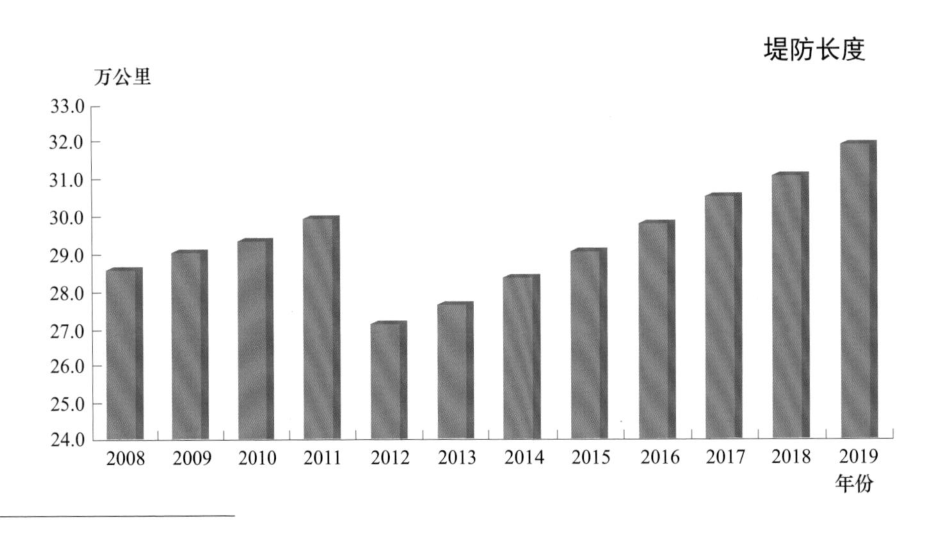

[1] 2011年以前各年堤防长度含部分地区5级以下江河堤防长度。

水库和枢纽。全国已建成各类水库98112座，水库总库容8983亿立方米。其中：大型水库744座，总库容7150亿立方米；中型水库3978座，总库容1127亿立方米。

机电井和泵站。全国已累计建成日取水大于等于20立方米的供水机电井或内径大于等于200毫米的灌溉机电井共511.7万眼。全国已建成各类装机流量1立方米每秒或装机功率50千瓦及以上的泵站96830处，其中：大型泵站383处，中型泵站4330处，小型泵站92117处。

灌区工程。全国已建成设计灌溉面积2000亩及以上的灌区共22844处，耕地灌溉面积37663千公顷。其中：50万亩及以上大型灌区176处，耕地灌溉面积12609千公顷；30万~50万亩大型灌区284处，耕地灌溉面积5386千公顷；1万~30万亩中型灌区7424处，耕地灌溉面积15506千公顷。截至2019年年底，全国灌溉面积75034千公顷，耕地灌溉面积68679千公顷，占全国耕地面积的51.0%。全国节水灌溉工程面积37059千公顷，高效节水灌溉面积22640千公顷，其中：喷灌、微灌面积11598千公顷，低压管灌面积11043千公顷。

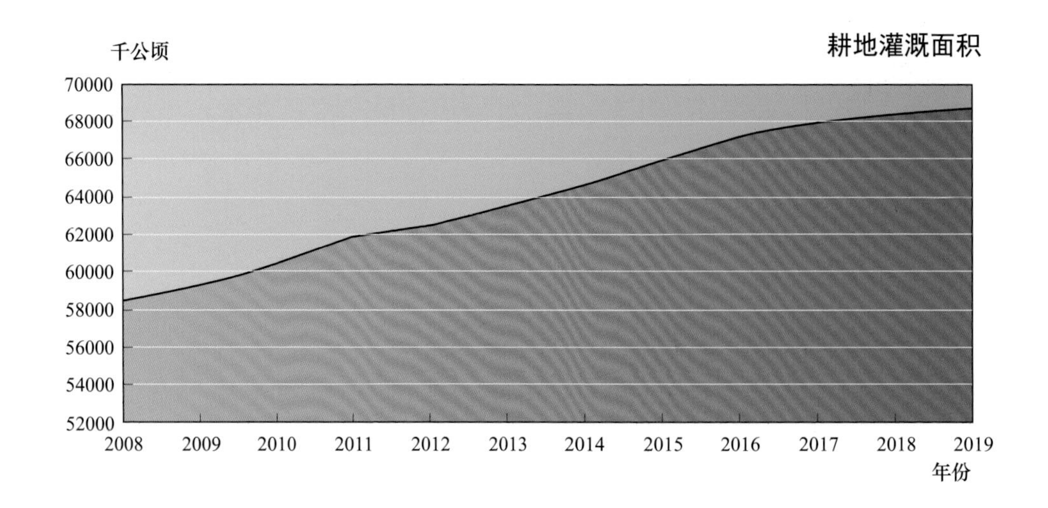

农村水电。截至2019年年底，全国已建成农村水电站45445座，装机容量8144.2万千瓦，占全国水电装机容量的22.9%；年发电量2533.2亿千瓦时，占全国水电发电量的19.5%。

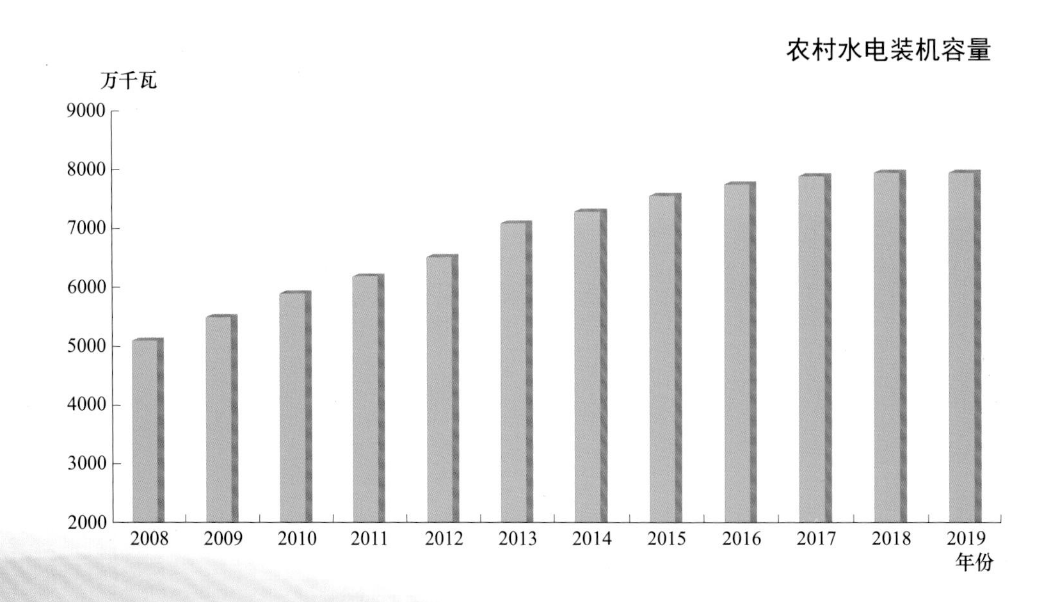

水土保持工程。全国累计水土流失综合治理面积达137.3万平方公里[1]，累计封禁治理保有面积达25.2万平方公里。2019年持续开展全国全覆盖的水土流失动态监测工作，全面掌握县级以上行政区、重点区域、大江大河流域的水土流失动态变化。

水土流失治理面积

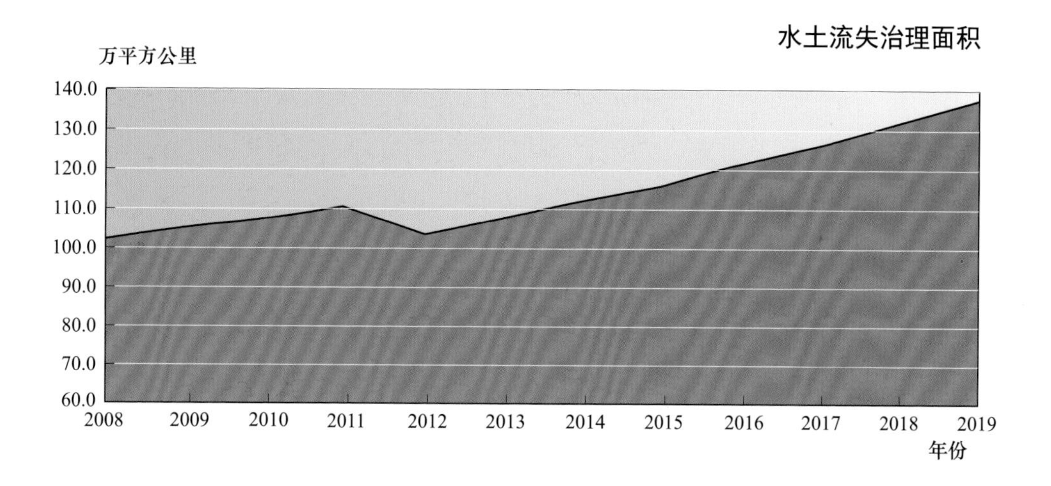

水文站网。全国已建成各类水文测站119608处，包括国家基本水文站3210处、专用水文站4435处、水位站15294处、雨量站53908处、蒸发站12处、地下水站26020处、水质站12712处、墒情站3961处、实验站56处。其中，向县级及以上水行政主管部门报送水文信息的各类水文测站66956处，可发布预报站2229处，可发布预警站1476处；配备在线测流系统的水文测站1693处，配备视频监控系统的水文测站3540处。基本建成由中央、流域、省级和地市级共333个水质监测（分）中心和水质站（断面）组成的水质监测体系。

[1] 2012年数据与第一次全国水利普查数据进行了衔接。

水利网信。截至2019年年底，全国省级以上水利部门配置累计各类服务器7476台（套），形成存储能力33.8PB，存储各类信息资源总量达2.1PB；县级以上水利部门累计配置各类卫星设备3696台（套），利用北斗卫星短文传输报汛站达6383多个，应急通信车65辆，集群通信终端3301个，宽、窄带单通信系统239套，无人机1008架。全国省级以上水利部门各类信息采集点达43.6万处，其中：水文、水资源、水土保持等各类采集点共约21.1万个，大中型水库安全监测采集点约22.5万个。

4 水资源节约利用与保护

水资源状况。2019 年，全国水资源总量 29041.0 亿立方米，比多年平均值偏多 4.8%；全国年平均降水量[1] 651.3 毫米，比多年平均偏多 1.4%，较上年减少 4.6%；截至 2019 年年底，全国 677 座大型水库和 3628 座中型水库年末蓄水总量 4118.4 亿立方米，比年初减少 91.7 亿立方米。

水资源开发。2019 年，新增规模以上水利工程[2]供水能力 68.0 亿立方米。截至 2019 年年底，全国水利工程供水能力达 8793.0 亿立方米，其中：跨县级区域供水工程 598.9 亿立方米，水库工程 2352.2 亿立方米，河湖引水工程 2101.6 亿立方米，河湖泵站工程 1804.0 亿立方米，机电井工程 1414.4 亿立方米，塘坝窖池工程 362.6 亿立方米，非常规水资源利用工程 159.3 亿立方米。

---

[1] 2019 年全国平均年降水量依据约 18000 个雨量站观测资料评价。

[2] 规模以上水利工程包括：总库容大于等于 10 万立方米的水库、装机流量大于等于 1 立方米每秒或装机容量大于等于 50 千瓦的河湖取水泵站、过闸流量大于等于 1 立方米每秒的河湖引水闸、井口井壁管内径大于等于 200 毫米的灌溉机电井和日供水量大于等于 20 立方米的机电井。

水资源利用。2019 年，全年总供水量为 6021.2 亿立方米，其中：地表水供水量 4982.5 亿立方米，地下水供水量 934.2 亿立方米，其他水源供水量 104.5 亿立方米。全国总用水量 6021.2 亿立方米，其中：生活用水 871.7 亿立方米，工业用水 1217.6 亿立方米，农业用水 3682.3 亿立方米，人工生态环境补水 249.6 亿立方米。与上年比较，用水量增加 5.7 亿立方米，其中：农业用水量减少 10.9 亿立方米，工业用水量减少 44.1 亿立方米，生活用水及人工生态环境补水量分别增加 11.9 亿立方米和 48.8 亿立方米。全国人均综合用水量为 431 立方米，农田灌溉水有效利用系数 0.559，万元国内生产总值（当年价）用水量 60.8 立方米，万元工业增加值（当年价）用水量 38.4 立方米。按可比价计算，万元国内生产总值用水量和万元工业增加值用水量分别比 2018 年下降 5.7%和 8.7%。

5 防洪抗旱

2019年，全国洪涝灾害总体偏轻，洪涝灾害直接经济损失1922.7亿元（水利设施直接损失409.4亿元），占当年GDP的0.19%。全国农作物受灾面积6680.4千公顷，绝收面积1321.5千公顷，受灾人口4766.6万人，因灾死亡573人、失踪85人，倒塌房屋10.3万间[1]。黑龙江、江西、山东、湖南、四川等地区受灾较重。全国因山洪灾害造成人员死亡347人，占全部死亡人数的60.6%。

全国受旱地域分布较广，但造成的影响总体较轻，江西、安徽、福建、湖北、湖南、重庆等省（直辖市）旱灾比较严重。全国农田因旱受灾面积8777.8千公顷，成灾面积4179.9千公顷，直接经济总损失376.8亿元。全国因旱累计有692.3万城乡人口、368.1万头大牲畜发生临时性饮水困难。

[1] 2019年洪涝灾害直接经济损失、全国农作物受灾面积、绝收面积、受灾人口、因灾死亡和失踪人口、倒塌房屋数量等数据来源于应急管理部国家减灾中心。

历年水旱灾害受灾、成灾情况

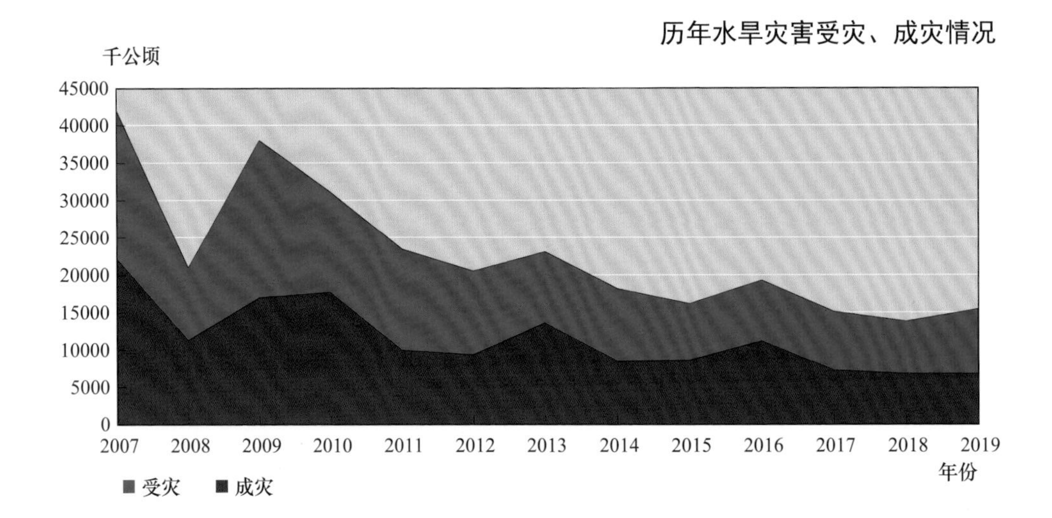

全年中央下拨水利救灾资金29.1亿元，其中：防汛资金19.4亿元，抗旱资金9.7亿元。全年完成抗旱浇地面积23553千公顷，抗旱挽回粮食损失249.1亿公斤，解决了669.8万城乡居民和289.7万头大牲畜因旱临时饮水困难。

6 水利改革与管理

节约用水管理。以县域为单元全面开展节水型社会达标建设，北京等19个省（自治区、直辖市）共计266个县级行政区通过达标验收。推广高校合同节水，40所高校签约实施，吸引社会资本约1.30亿元，平均节水率约为20%。开展水利行业节水机关建设，43个部属单位（含部机关）和省厅建成节水型机关，预计年均节水量达27.1万立方米，平均节水率29%。

河（湖）长制。2019年，31个省（自治区、直辖市）党委政府主要负责同志全部担任总河长，明确省、市、县、乡级河长湖长30多万名、村级河长湖长（含巡河员、护河员）90多万名，实现河湖管护责任全覆盖。全国省、市、县、乡级河长湖长累计巡查河湖712万人次，组织实施“一河（湖）一策”。组织开展河湖“清四乱”专项行动，全国累计清理整治河湖“四乱”问题13.7万个，清除河道垃圾3500万吨，拆除违法建筑面积3900万平方米，清理非法占用河湖岸线2.5万公里，拆除非法围堤9800公里，清除非法围网面积8600万平方米，清除河道非法采砂点1.04万个，查处非法采运砂船9100艘，河湖面貌明显好转。组织开展长江干流岸线利用项目整治专项行动，共排查出2441个涉嫌违法违规项目，完成整治2230个；组织开展长江经济带固体废物清理整治专项行动，河湖管理范围内1376处固体废物全部清零；与交通运输部、公安部联合开展长江干流采砂专项整治、清江行动等。强化监督检查，对31个省（自治区、直辖市）所有设区市的6679条河流（段）、1612个湖泊开展暗访督查；对长江、黄河、大运河等重点流域区域开展专项督查。

最严格水资源管理。2019年，水利部会同国家发展改革委等完成对31个省（自治区、直辖市）2019年度最严格水资源管理制度考核，江苏等8个省（自治区、直辖市）考核等级为优秀。无定河等5条跨省江河流域水量分配方案得以批复，湖南等16个省（自治区、直辖市）开展了省内跨市县江河水量分配工作。2018—2019调水年度南水北调东线一期工程向山东调水8.44亿立方米，中线一期工程向北京、天津、河北、河南4省（直辖市）调水共计71.32亿立方米，受水区供水安全保障能力显著提升。加强重要跨省江河流域水资源统一调度，汉江、嘉陵江、乌江等15条跨省江河实施统一调度，黄河干流实现连续20年不断流，黑河下游东居延海连续15年不干涸。105个水生态文明试点城市建设进展顺利，其中，99个试点城市通过验收。实施华北地下水超采综合治理河湖地下水回补试点工作，利用南水北调中线工程和河北省水库对河北省3条河流进行地下水回补，全年补水9.49亿立方米。中国水权交易所2019年完成水权交易237单，交易水量1.15亿立方米。

运行管理。2019 年，落实小型水库维修养护中央补助资金 14.5 亿元，带动地方财政投入 8.9 亿元。培训小型水库管理人员和巡查管护人员 9.3 万人次。截至 2019 年年底，累计批准国家级水利风景区 878 个，其中：水库型 373 个，自然河湖型 195 个，城市河湖型 195 个，湿地型 47 个，灌区型 31 个，水土保持型 37 个。

水价改革。组织水利工程供水价格调查，开展水价关键问题研究。推动国家发展改革委完善水价管理制度，修订出台《中央定价目录》，将价格偏低、长期未做调整的部属工程列入成本监审计划。截至 2019 年年底，累计实施农业水价综合改革面积 2.9 亿亩，其中 2019 年新增农业水价综合改革面积 1.3 亿亩。

水利规划和前期工作。2019 年，中央层面审批（含印发审查意见）水利规划 14 项。坚持开门编规划，采取网上公开征集，部署开展“十四五”水安全保障规划编制工作。经国务院同意，会同财政部、国家发展改革委、农业农村部联合印发《华北地区地下水超采综合治理行动方案》。扎实推进国家重大战略水利规划工作，全力抓好京津冀协同发展、推动长江经济带发展、长江三角洲区域一体化发展、粤港澳大湾区建设、黄河流域生态保护和高质量发展、推进海南全面深化改革开放等国家重大区域发展战略水利各项任务落实。加快推进重点流域和主要支流综合规划审批，批复洮河、伊洛河、窟野河、湘江、资水等流域综合规划。加强水利与国土空间总体规划衔接，启动水利基础设施空间布局规划编制工作。国家发展改革委批复项目前期工作 14

项，其中可行性研究报告13项、工程规划1项，总投资411.47亿元。水利部批复初步设计12项，总投资957.79亿元。

水土保持管理。2019年，水利部制定出台了《进一步深化“放管服”改革全面加强水土保持监管的意见》等13项制度。全国共审批生产建设项目水土保持方案5.59万个，涉及水土流失防治责任范围2.08万平方公里；1.36万个生产建设项目完成水土保持设施自主验收报备。以创新手段推行生产建设项目人为水土流失遥感监管，范围覆盖约592万平方公里国土面积，通过卫星遥感解译组织现场复核，共认定并查处“未批先建”“未批先弃”等违法违规项目5.3万个。

农村水电管理。截至2019年年底，21个省份累计创建绿色小水电示范电站338座。积极推进农村水电站安全生产标准化建设，全国累计建成安全生产标准化电站2789座，其中一级电站82座、二级电站1168座、三级电站1539座。全国共有1168条河流、1869个生态改造项目、1997个增效扩容项目完成改造，累计修复减脱水河段2500多公里。

水利移民。2019年，搬迁人口21.7万人，其中：农村移民搬迁19.2万人，城集镇移民搬迁2.5万人，生产安置17.3万人。国家核定上一年度新建大中型水库农村移民后期扶持人数11.3万人。

水利监督。2019 年，组织开展了全国河湖长制落实情况监督检查、2019 年度最严格水资源管理制度考核、水利行业“强监管”检查等三大类督查检查考核，全年共派出督查检查组 2091 组次、8412 人次，检查 31049 个项目，发现问题 49629 项；采取“四不两直”方式检查的项目 1741 组次，占总派出组次的 83.3%，共发出“约谈”及以上等级责任追究文件 45 份，对 759 家责任单位实施了 834 家・次“约谈”及以上等级责任追究。水利行业共发生 8 起生产安全事故，死亡 11 人。开展水利工程建设安全生产专项整治、水利行业安全生产集中整治、水利工程建设安全生产巡查和重大水利工程质量与安全巡查等，共排查出各类隐患 47236 个，整改率达到 97.8%。探索“安全监管+信息化”模式，4.8 万个单位纳入安全监管平台；103 家单位通过水利安全生产标准化达标审查，64 家部直属单位完成标准化达标创建；出台了《水利水电工程（水库、水闸）运行危险源辨识与风险评价导则（试行）》。全年开展稽查 7 批次，共派出稽查组 100 个、815 人次，涉及项目 252 个，共发现问题 6000 余个，印发“一省一单”整改意见通知 99 份。推动省级水行政主管部门开展自主稽查 170 余批次，共派出稽查组 500 余个，近 3000 人次，涉及项目近 2000 个，共发现问题 16000 余个。

依法行政。2019 年，修订水利部规章 4 件，修订规范性文件 2 件，废止 1 件。全国立案查处水事违法案件 25115 件，结案 22927 件，结案率 91.3%；水利部共办结行政复议案件 23 件，办理行政应诉 7 件。

行政许可。2019年，水利部（包括部机关和各流域机构）共受理行政审批事项1667件，办结1328件。其中：水工程建设规划同意书审核19件，水利基建项目初步设计文件审批12件，取水许可发放403件，非防洪建设项目洪水影响评价报告审批33件，河道管理范围内建设项目工程建设方案审批282件，生产建设项目水土保持方案审批90件，国家基本水文测站设立和调整审批27件，国家基本水文测站上下游建设影响水文监测工程的审批27件，水利工程建设监理单位资质认定（新申请、增项、晋升、延续）95件，水利工程质量检测单位甲级资质认定（新申请、增项、晋升、延续）360件。

水利科技。2019年，国家立项安排1.01亿元资金用于水利科技项目，其中：组织承担国家重点研发计划“水资源高效开发利用”和“重大自然灾害监测预警与防范”等涉水重点专项项目共3项，合计6273万元；水利技术示范项目57项，合计3812.83万元。水利部、国家自然科学基金委员会和中国长江三峡集团共同设立长江科学研究联合基金，落实经费25000万元。组织开展水利重大科技问题研究21项。水利科技成果获得国家科技进步奖特等奖1项、二等奖2项。截至2019年年底，水利系统共有国家和部级重点实验室12个，国家和部级工程技术研究中心15个，部级野外科学观测研究站6个。落实中央财政公益性科研院所基本科研业务费13985.4万元。发布水利技术标准33项，在编141项。截至2019年年底，水利行业现行有效标准达854项。

国际合作。2019 年，共签署水利国际合作协议 6 份，在华举办多边、双边高层圆桌会议或技术交流研讨会 13 次。亚洲开发银行、全球环境基金开展的 5 个项目进展顺利，中瑞（士）、中丹、中法、中芬、中荷合作项目、国际科技合作的援外项目稳步开展。

7 水利行业状况

水利单位。截至2019年年底，水利系统内外各类县级及以上独立核算的法人单位23554个，从业人员93.8万人。其中：机关单位2728个，从业人员12.4万人，比上年减少1.6%；事业单位16145个，从业人员50.0万人，比上年减少15.0%；企业3758个，从业人员30.6万人，比上年减少11.0%；社团及其他组织923个，从业人员0.7万人，比上年减少56.3%。全国共有水利水电工程施工总承包特级资质企业27家、一级资质企业264家。

职工与工资。全国水利系统从业人员85.4万人，比上年减少5.4%。其中：全国水利系统在岗职工82.7万人，比上年减少5.9%。在岗职工中，部直属单位在岗职工6.6万人，比上年增加0.6%；地方水利系统在岗职工76万人，比上年减少6.5%。全国水利系统在岗职工工资总额为787.6亿元，年平均工资9.5万元。

职工与工资情况

	2009年	2010年	2011年	2012年	2013年	2014年	2015年	2016年	2017年	2018年	2019年
在岗职工人数/万人	103.7	106.6	102.5	103.4	100.5	97.1	94.7	92.5	90.4	87.9	82.7
其中：部直属单位/万人	7.2	7.4	7.5	7.4	7.0	6.7	6.6	6.4	6.4	6.6	6.6
地方水利系统/万人	96.5	96.3	95.0	96.0	93.5	90.4	88.1	86.1	84.0	81.3	76.0
在岗职工工资/亿元	264.7	297.9	351.4	389.1	415.3	451.4	529.4	640.5	739.1	802.7	787.6
年平均工资/(元/人)	25633	28816	34283	37692	41453	46569	55870	69377	83534	91307	95385

全国水利发展主要指标（2014—2019年）

指标名称	单位	2014年	2015年	2016年	2017年	2018年	2019年
1. 灌溉面积	千公顷	70652	72061	73177	73946	74542	75034
2. 耕地灌溉面积	千公顷	64540	65873	67141	67816	68272	68679
其中：本年新增	千公顷	1648	1798	1561	1070	828	780
3. 节水灌溉面积	千公顷	29019	31060	32847	34319	36135	37059
其中：高效节水灌溉面积	千公顷	16114	17923	19405	20551	21903	22640
4. 万亩以上灌区	处	7709	7773	7806	7839	7881	7884
其中：30万亩以上	处	456	456	458	458	461	460
万亩以上灌区耕地灌溉面积	千公顷	30256	32302	33045	33262	33324	33501
其中：30万亩以上	千公顷	11251	17686	17765	17840	17799	17994
5. 自来水普及率	%		76	79	80	81	82
农村集中供水率	%		82	84	85	86	87
6. 除涝面积	千公顷	22369	22713	23067	23824	24262	24530
7. 水土流失治理面积	万平方公里	111.6	115.5	120.4	125.8	131.5	137.3
其中：新增	万平方公里	5.5	5.4	5.6	5.9	6.4	6.7
8. 水库	座	97735	97988	98460	98795	98822	98112
其中：大型水库	座	697	707	720	732	736	744
中型水库	座	3799	3844	3890	3934	3954	3978
水库总库容	亿立方米	8394	8581	8967	9035	8953	8983
其中：大型水库	亿立方米	6617	6812	7166	7210	7117	7150
中型水库	亿立方米	1075	1068	1096	1117	1126	1127
9. 全年水利工程总供水量	亿立方米	6095	6103	6040	6043	6039	6021
10. 堤防长度	万公里	28.4	29.1	29.9	30.6	31.2	32.0
保护耕地	千公顷	42794	40844	41087	40946	41351	41903
堤防保护人口	万人	58584	58608	59468	60557	62785	64168
11. 水闸总计	座	98686	103964	105283	103878	104403	103575
其中：大型水闸	座	875	888	892	892	897	892

续表

指标名称	单位	2014年	2015年	2016年	2017年	2018年	2019年
12. 年末全国水电装机容量	万千瓦	30183	31937	33153	34168	35226	35564
全年发电量	亿千瓦时	10661	11143	11815	11967	12329	12991
13. 农村水电装机容量	万千瓦	7322	7583	7791	7927	8044	8144
全年发电量	亿千瓦时	2281	2351	2682	2477	2346	2533
14. 当年完成水利建设投资	亿元	4083.1	5452.2	6099.6	7132.4	6602.6	6711.7
按投资来源分：							
（1）中央政府投资	亿元	1648.5	2231.2	1679.2	1757.1	1752.7	1751.1
（2）地方政府投资	亿元	1862.5	2554.6	2898.2	3578.2	3259.6	3487.9
（3）国内贷款	亿元	299.6	338.6	879.6	925.8	752.5	636.3
（4）利用外资	亿元	4.3	7.6	7.0	8.0	4.9	5.7
（5）企业和私人投资	亿元	89.9	187.9	424.7	600.8	565.1	588.0
（6）债券	亿元	1.7	0.4	3.8	26.5	41.6	10.0
（7）其他投资	亿元	176.5	131.7	207.1	235.9	226.3	232.8
按投资用途分：							
（1）防洪工程	亿元	1522.6	1930.3	2077.0	2438.8	2175.4	2289.8
（2）水资源工程	亿元	1852.2	2708.3	2585.2	2704.9	2550.0	2448.3
（3）水土保持及生态建设	亿元	141.3	192.9	403.7	682.6	741.4	913.4
（4）水电工程	亿元	216.9	152.1	166.6	145.8	121.0	106.7
（5）行业能力建设	亿元	40.9	29.2	56.9	31.5	47.0	63.4
（6）前期工作	亿元	65.1	101.9	174.0	181.2	132.0	132.7
（7）其他	亿元	244.2	337.5	636.2	947.5	835.8	757.4

说明：1. 本公报不包括香港特别行政区、澳门特别行政区及台湾省的数据。
2. 水利发展主要指标分别于2012年、2013年与第一次全国水利普查数据进行了衔接。
3. 农村水电的统计口径为单站装机容量5万千瓦及以下的水电站。

2019 STATISTIC BULLETIN ON CHINA WATER ACTIVITIES

Ministry of Water Resources, P. R. China

The year of 2019 marks the 70th anniversary of the founding of the People's Republic of China. It is the critical year for the building of a moderately prosperous society in all respects and also the first year to implement the keynote for water sector reform and development. The Party Central Committee and the State Council pay high attention to water resources management. Over the past year, under the guidance of Xi Jinping's thought of socialism with Chinese characteristics for a new era, water departments and authorities at all levels had made great efforts to implement all decisions and deployments of the CPC Central Committee and the State Council, by strengthening leadership with political construction, sticking to the keynote for water sector reform and development, targeting goals, problems, and results with practical measures, and made great achievements in water resources management despite all the difficulties.

I. Investment in Fixed Assets

In 2019, the total investment in water projects amounted to 671. 17 billion Yuan, with an increase of 10. 91 billion or 1. 7% compared with that in 2018. In which, 498. 79 billion Yuan was being allocated for construction projects with an increase of 2. 3%, 24. 31 billion Yuan for installation with a decrease of 13. 5%, 22. 11 billion Yuan for expenditure on purchases of machinery, electric equipment and instruments with an increase of 3. 1%, and 125. 97 billion Yuan for other purposes, including compensation for resettlement and land acquisition, with an increase of 2. 4%.

	2012 /billion Yuan	2013 /billion Yuan	2014 /billion Yuan	2015 /billion Yuan	2016 /billion Yuan	2017 /billion Yuan	2018 /billion Yuan	2019 /billion Yuan	Increase /%
Total completed investment	396. 42	375. 76	408. 31	545. 22	609. 96	713. 24	660. 26	671. 17	1. 7
Construction project	273. 65	278. 28	308. 64	415. 08	442. 20	506. 97	487. 72	498. 79	2. 3
Installation project	23. 78	17. 36	18. 50	22. 88	25. 45	26. 58	28. 09	24. 31	-13. 5
Purchase of machinery, equipment and instruments	17. 81	16. 11	20. 61	19. 87	17. 28	21. 17	21. 44	22. 11	3. 1
Others (including compensation of resettlement and land acquisition)	81. 18	64. 02	60. 56	87. 39	125. 03	158. 52	123. 01	125. 97	2. 4

In the total completed investment, 228. 98 billion Yuan was allocated to the construction of flood control projects, up 5. 3% from the previous year; 244. 83 billion Yuan was for the construction of water resources projects, down 4. 0%; 91. 34 billion Yuan was for soil and water conservation and ecological restoration, up 23. 2%; and 106. 02 billion Yuan for specific projects of hydropower development and capacity building, down by 6. 7% over the previous year.

The competed investment for seven major river basins reached 526. 21 billion Yuan, of which 144. 96 billion Yuan was invested in river basins in the southeast, southwest and northwest of China, while investments in river basins in east, central, west and northeast China were 259. 34 billion Yuan, 165. 32 billion Yuan, 224. 52 billion Yuan and 21. 99 billion Yuan, respectively.

Completed investment of projects in 2019

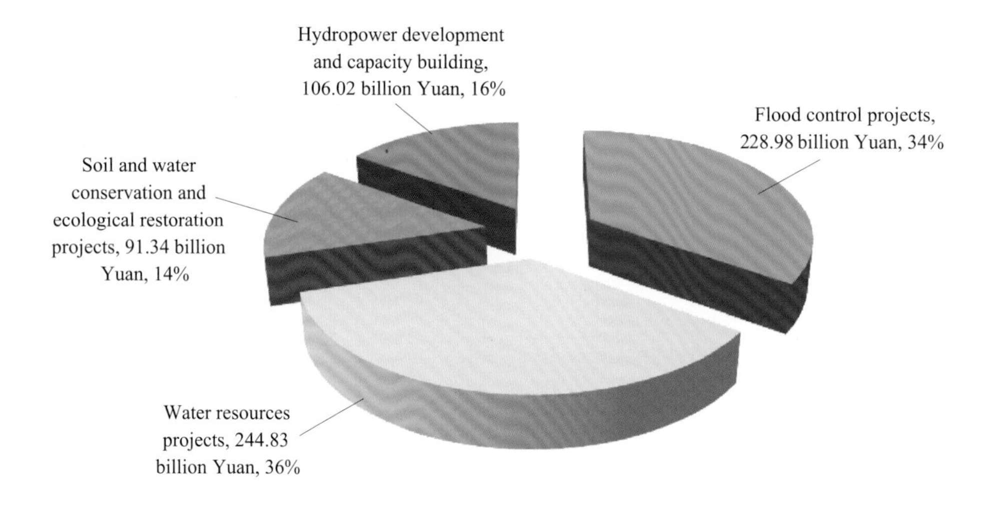

Of the total competed investment, the Central Government contributed 6. 64 billion Yuan, and local governments contributed 664. 53 billion Yuan. Investments for large and medium-sized projects were 106. 68 billion Yuan; and for small and other projects were 564. 49 billion Yuan. Investments for new projects and rehabilitation and expansion projects were 520. 28 billion Yuan and 150. 89 billion Yuan.

The newly-added fixed asset of the year in water project construction totaled 347. 38 billion Yuan. By the end of 2019, the accumulated investment in projects under construction was 1, 764. 94 billion Yuan, with the completion rate reaching 62. 7%. New fixed assets totaled 998. 76 billion Yuan and the rate of investment transferred into fixed assets was 56. 6%, an increase of 0. 1 percentage point over the previous year.

A total of 28, 742 water projects were under construction in 2019, with a total investment of 2, 816. 69 billion Yuan, an increase of 2. 4% over the previous year. Among which, 17, 202 projects were funded by the Central Government, an increase of 1. 6% over the previous year, occupying 1, 405. 83 billion Yuan, down 1. 0% over the previous year. There were 20, 779 projects launched in 2019, an increase of 5. 0%, with an increase of investment totaled 686. 80 billion Yuan, an increase of 13. 3% over the previous year. The completed civil works of earth, stone and concrete structures were 2. 99 billion m^3, 390 million m^3, and 190 million m^3, respectively. By the end of 2019, the rate of completed quantity of earthwork, stonework, and concrete of under-construction projects were 98. 7%, 94. 3% and 95. 5%, respectively.

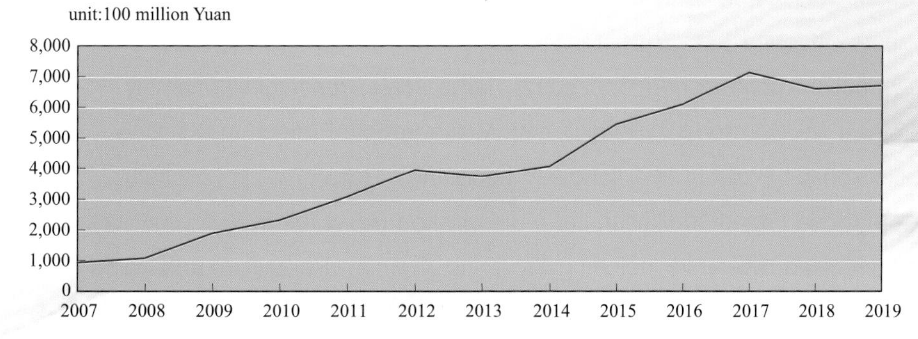

II. Key Water Projects Construction

Harness of large rivers and lakes. In 2019, there were 5, 414 river harness projects under construction, including 601 flood control dyke and embankment construction projects, 739 projects for large river and main tributary control, and 3, 331 medium-sized and small river control works, 743 flood diversion and storage area construction or other projects. By the end of 2019, the accumulated investment in projects under construction was 388. 29 billion Yuan, with a completion rate of 67. 2%. The projects for river regime control and river course training and restoration in the middle and lower reaches of the Yangtze River had been effectively implemented. The flood control works in the lower reaches of the Yellow River were completed in the main. Out of the 38 Huaihe River improvement projects, 32 started construction, among which 8 were put into operation for benefit generation. Improvement work for Dongting and Poyang lakes also made significant progress. Up to 19 out of the 21 projects for the Comprehensive Improvement of Water Environment of Taihu Lake began construction, among which 10 projects completed construction and created benefits.

Reservoir projects. In 2019, there were 1, 360 reservoir projects under construction. By the end of the year, completed investment of under-construction projects reached 319. 75 billion Yuan, with a completion rate of 60. 3%. Construction commenced for some large-scale reservoirs, namely Fengshan Reservoir in Guizhou, Yulong Kashi Multipurpose Project in Xinjiang, Huaqiao Multipurpose Project in Jiangxi, Longtang Reservoir in Sichuan, Maiwan Multipurpose Project in Hainan, and Jiangjiakou Reservoir in Sichuan, as well as 5 medium-sized reservoirs including Xiyin and Xiushui. Reservoirs and dams such as Datengxia, Dashixia in Xinjiang, Nianpanshan in Hubei, Tuxikou in Sichuan, and Lijiayan in Sichuan all achieved their annual targets of water diversion and damming. Reservoirs were impounded, such as Chushandian Reservoir in Henan,

Aertashi in Xinjiang, Laluo in Tibet, Shuangfengsi Reservoir in Hebei, Zhuangli Reservoir in Shandong, Mangshan Reservoir in Hunan, Fendou Reservoir in Heilongjiang, and Maling Reservoir in Guizhou.

Water allocation projects. In 2019, investment in water allocation projects under construction reached 787. 5 billion Yuan and completed investment accumulated to 453. 02 billion Yuan, accounting for 57. 5% of the total. Water allocation to Pearl River Delta in Guangdong started construction. Water transfers were started for allocating water from Fengjiangkou Reservoir in the upstream to the northern Hubei and Phase-II Irrigation Scheme of the Liaofang Multipurpose Project in Jiangxi. Progressed sped up for the projects of water diversion from the Yangtze River to the Huaihe River, water diversion in central Yunnan, water diversion from the Chuoer River to the Xiliao River in Inner Mongolia, water diversion from the Hanshui River to the Weihe River in Shaanxi, Jiayan Multipurpose Project in Guizhou, and water diversion in northwestern Guizhou.

Irrigation, drainage and rural water supply. In 2019, completed investment for strengthening and improving safe drinking water supply in rural areas reached 61. 96 billion Yuan, including 12. 17 billion Yuan of central government subsidies. The Central Government allocated 8. 6 billion Yuan for the construction of large and medium irrigation and drainage systems and rehabilitation of irrigation districts for water saving purpose, 6. 35 billion Yuan for water conservation in key medium-sized irrigation districts, and 7. 96 billion Yuan for the construction of highly-efficient water-saving farmland waterworks. The effective irrigated area increased by 780, 000 ha, water-saving irrigated area increased by 1, 076, 000 ha, and area covered by highly-efficient water-saving irrigation grew by 756, 000 ha. By the end of 2019, the rural population access to tap water supply made up a percentage of 82% and the coverage of centralized water supply raised to 87%.

Rural hydropower and electrification. In 2019, completed investment of rural hydropower station construction nationwide amounted to 7.1 billion Yuan, building 90 new hydropower stations, with a total installed capacity of 1.072 million kW. Of these, newly installed capacity amounted to 0.564 million kW, and capacity increase resulted from rehabilitation was 0.508 million kW.

Soil and water conservation. In 2019, a total of 109.08 billion Yuan was allocated to the under-constructed projects for soil and water conservation and ecological restoration, with an accumulated investment of 91.34 billion Yuan. The newly-added areas for comprehensive control of soil erosion reached 67,000 km^2, of which the areas under the National Major Project for Soil Conservation was 11,000 km^2. Up to 600 silt-retention dams on Loess Plateau at high risk were strengthened and rehabilitated.

Capacity building. The completed investment for capacity building in 2019 was 11.49 billion Yuan, of which 850 million Yuan was spent on communication equipment for flood control, 1.88 billion Yuan on hydrological facilities, 160 million Yuan for scientific research and education facilities and 8.6 billion Yuan for others.

III. Key Water Facilities

Embankments and water gates. By the end of 2019, the completed river dykes and embankments at Grade-V or above had a total length of 320, 000 km❶. The accumulated length of dykes and embankments that met the standard reached 227, 000 km, accounting for 71. 0% of the total. Specifically, Grade-I and Grade-II dykes and embankments up to the standard reached 35, 000 km, or 81. 7% of the total. All completed dykes and embankments nationwide can protect 640 million people and 42, 000 ha of cultivated land. The number of water gates with a flow of 5 m^3/s increased to 103, 575, of which 892 were large water gates. By type, there were 8, 293 flood diversion sluices, 18, 449 drainage/return water sluices, 5, 172 tidal barrages, 13, 830 water diversion intakes and 57, 831 controlling gates.

Length of embankment

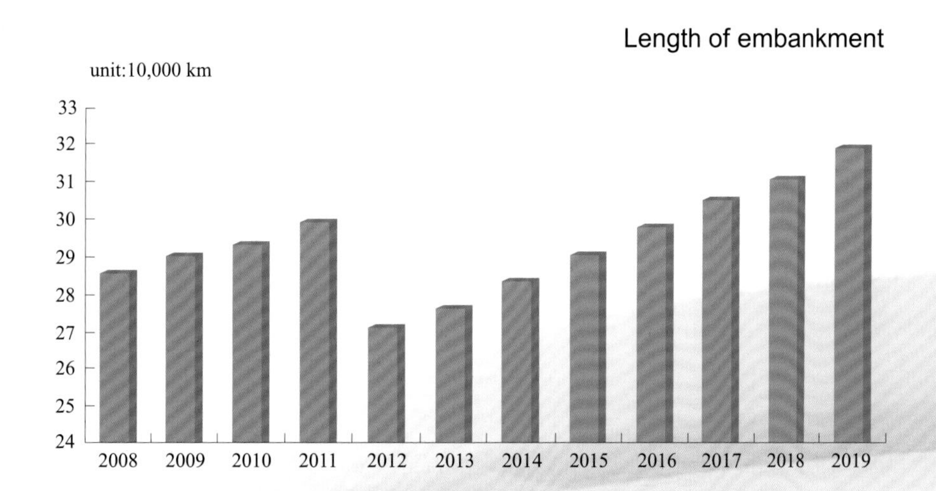

❶ The length of embankment before 2011 includes embankment below Grade-V.

Reservoirs and water complexes. The number of reservoirs in China reached 98, 112, with a total storage capacity of 898. 3 billion m^3. Of which 744 reservoirs are large reservoirs, with a total capacity of 715 billion m^3 and 3, 978 reservoirs are medium-sized once with a total capacity of 112. 7 billion m^3.

Tube wells and pumping stations. Accumulatively, a total of 5. 117 million tube wells, with a daily water abstraction capacity equal or larger than 20 m^3 or an inner diameter larger than 200 mm, had been completed for water supply in the whole country. A total of 96, 830 pumping stations with a flow of 1 m^3/s or installed voltage above 50 kW were put into operation, including 383 larger, 4, 330 medium-size and 92, 117 small-size pumping stations.

Irrigation systems. The irrigation districts with an area of 2, 000 mu or above were 22, 844 in total, covering 37. 663 million ha of irrigated farmland. Of which, 176 irrigation districts had an irrigated area of 500, 000 mu or above, and their total irrigated area reached 12. 609 million ha. The irrigation districts with an area from 300, 000 to 500, 000 mu were 284, covering 5. 386 million ha of irrigated farmland. The irrigation districts with an area between 10, 000 and 300, 000 mu were 7, 424, covering 15. 506 million ha of irrigated farmland. By the end of 2019, the total irrigated area nationwide amounted to 75. 034 million ha; the irrigated area of cultivated land reached 68. 679 million ha that accounted for 51. 0% of the total in China. The areas with water-saving irrigation facilities totaled 37. 059 million ha, high efficiency water-saving irrigation totaled 22. 640 million ha, among which 11. 598 million ha were equipped with sprinklers or micro-irrigation systems and 11. 043 million ha had low-pressure pipes.

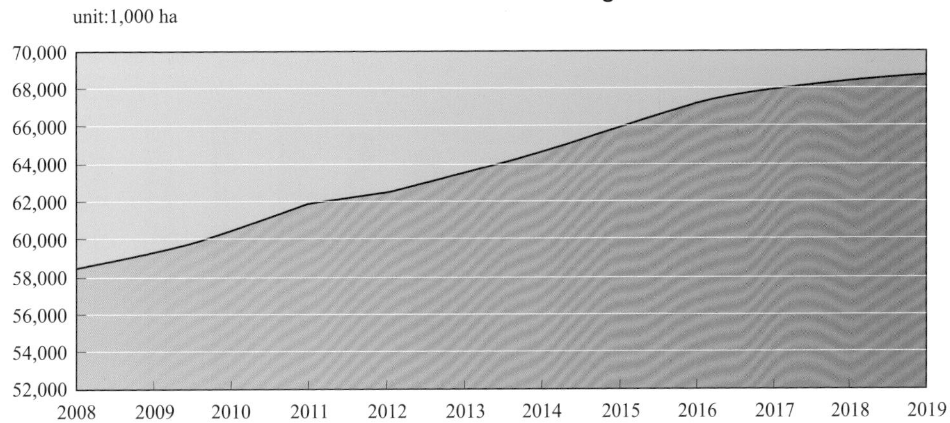

Rural hydropower and electrification. By the end of 2019, hydropower stations built in rural areas totaled 45, 445, with an installed capacity of 81. 442 million kW, accounting for 22. 9% of the national total. The annual power generation by these hydropower stations reached 253. 32 billion kW · h, accounting for 19. 5% of the national total.

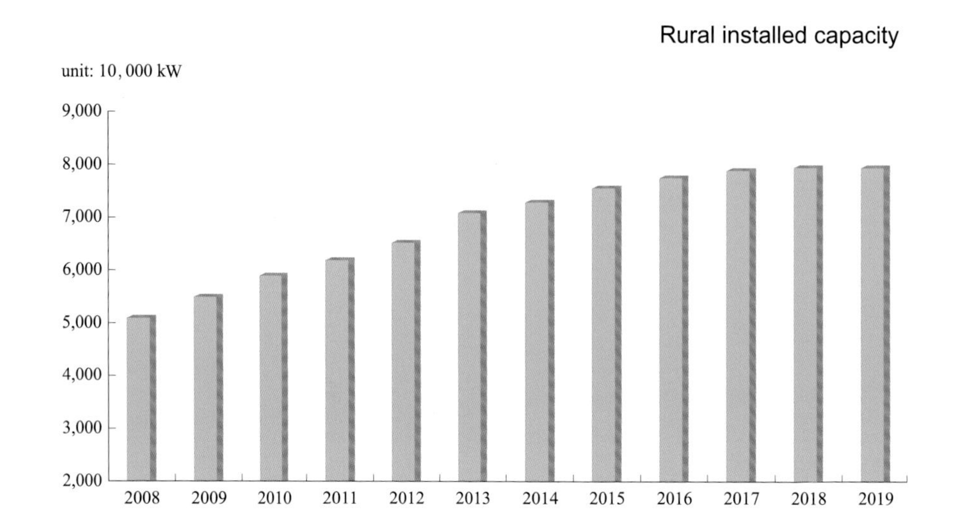

Soil and water conservation. By the end of 2019, the restored eroded areas reached 1.373 million km^2;❶ and the forbidden area for ecological restoration accumulated to 252,000 km^2. Dynamic monitoring for soil and water loss had been continued in all administrative areas above county level, key areas, and major river basins in the country, to gain a comprehensive understanding of dynamic changes.

---

❶ Statistical data in 2012 is integrated with the data of first national census for water.

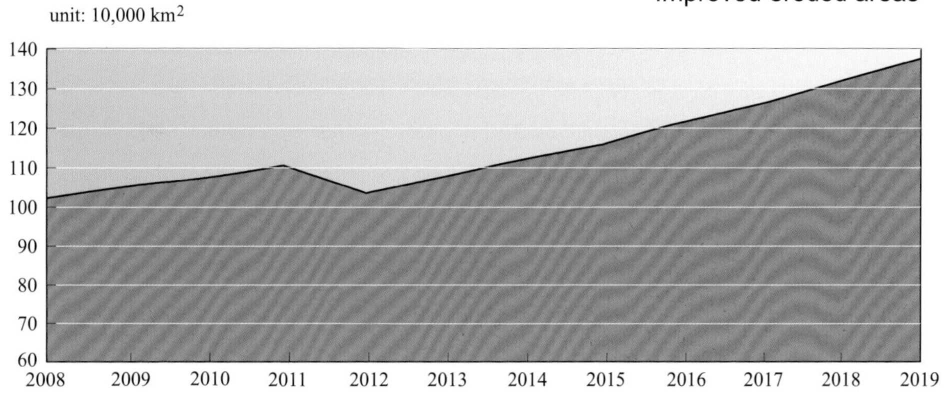

Hydrological station networks. In 2019, the number of hydrological stations of all kinds totaled 119, 608 in the whole country, including 3, 210 national basic hydrologic stations, 4, 435 special hydrologic stations, 15, 294 gauging stations, 53, 908 precipitation stations, 12 evaporation stations, 26, 020 groundwater monitoring stations, 12, 712 water quality stations, 3, 961 soil moisture monitoring stations and 56 experimental stations. Among which, 66, 956 stations of various kinds provided hydrological information to water administration authorities at and above county level; 2, 229 stations were employed for flood forecasting and 1, 476 for early warnings; 1, 693 were equipped for online flow measurement and 3, 540 had video monitors. A water quality monitoring system, including 333 monitoring centers and sub-centers as well as water quality stations (sections) at central, basin, provincial and local levels, had been formed.

Water networks and information systems. By the end of 2019, the water resources departments and authorities at and above provincial level were equipped with 7, 476 servers of varied kinds, forming a total storage capacity of 33. 8 PB, and keeping 2. 14 PB of data and information. The water resources departments and authorities at and above county level had equipped with 3, 696 sets of various kinds of satellite equipment, 6, 383 flood forecasting stations for short message transmission from the Beidou Satellites, and 65 vehicles for emergency communication, 3, 301 cluster communication terminals, 239 narrowband and broadband communication systems, and 1, 008 Unmanned Aerial Vehicles (UAV). A total of 436, 000 information gathering points were available for water resources departments and authorities at and above county level, including 211, 000 points for collecting data of hydrology, water resources and soil and water conservation and 225, 000 points for safety monitoring at large and medium-sized reservoirs.

IV. Water Resources Conservation, Utilization and Protection

Water resources conditions. The total national water resources in 2019 was 2904. 1 billion m^3, approximately 4. 8% more than the normal years. The mean annual precipitation❶ was 651. 3 mm, 1. 4% more than normal years and 4. 6% less than the previous year. By the end of 2019, the total storage of 677 large and 3, 628 medium-sized reservoirs were 411. 84 billion m^3, 9. 17 billion m^3 less than that at the beginning of the year.

Water resources development. In 2019, the newly-increased water supply capacity by water facilities above designated size❷ was 6. 8 billion m^3. By the end of 2019, the total water supply capacity of China reached 879. 30 billion m^3, including 59. 89 billion m^3 of water diverted from cross-county facilities, 235. 22 billion m^3 from reservoirs, 210. 16 billion m^3 from river and lake diversion schemes, 180. 40 billion m^3 from pumping stations along rivers and lakes, 141. 44 billion m^3 from tube wells, 36. 26 billion m^3 from ponds, weirs and cellars, and 15. 93 billion m^3 from unconventional water sources.

Water resources utilization. In 2019, the total water supply amounted to 602. 12 billion m^3, including 498. 25 billion m^3 from surface water, 93. 42 billion m^3 from groundwater and 10. 45 billion m^3 from other sources. The total water consumption amounted to 602. 12 billion m^3, among which domestic water use amounted to 87. 17 billion m^3, industrial water use totaled 121. 76 billion m^3, agricultural water

---

❶ Average precipitation of 2019 was based on data from approximately 18, 000 stations.

❷ Water projects above designated size include: reservoirs with a total capacity of 100, 000 m^3 or higher, pump stations with an installed flow at or above 1 m^3/s or an installed capacity at or above 50 kW, water diversion gates with a flow at or above 1 m^3/s, electric irrigation wells 200 mm or larger in inner diameter or with a water supply capacity at or above 20 m^3 per day.

use was 368. 23 billion m^3, artificial recharge for environmental and ecological use 24. 96 billion m^3. Comparing to the previous year, the total water consumption increased by 570 million m^3. Agricultural water use decreased by 1. 09 billion m^3 and industrial water use decreased by 4. 41 billion m^3. Nevertheless, domestic water use and artificial recharge for environmental and ecological purposes increased by 1. 19 billion m^3 and 4. 88 billion m^3 respectively. Water consumption per capita in 2019 was 431 m^3 in average. The coefficient of effective irrigated water use was 0. 559. Water use per 10, 000 Yuan of GDP (at comparable price of the same year) was 60. 8 m^3 and that per 10, 000 Yuan of industrial value added (at comparable price of the same year) was 38. 4 m^3. Based on estimation at comparable prices, water uses per 10, 000 Yuan of GDP and per 10, 000 Yuan of industrial value added decreased by 5. 7% and 8. 7% over the previous year respectively.

V. Flood Control and Drought Relief

In 2019, the overall damage caused by flood and water-logging disasters was less severe than other years on the whole, and the direct economic loss from flood disasters was 192. 27 billion Yuan (including 40. 94 billion Yuan of direct losses caused by damage to water facilities), accounting for 0. 19% of GDP in the same year. A total of 6, 680, 400 ha of cultivated land were affected by floods, 1, 321, 500 ha of farmland had no harvest, 47. 666 million people were affected with 573 dead and 85 missing. A total of 103, 000 houses collapsed[1]. Provinces suffered heavily from severe flooding included Heilongjiang, Jiangxi, Shandong, Hunan and Sichuan. Mountain flood took 347 lives, accounting to 60. 6% of the total death toll.

---

[1] Data for these indicators came from National Disaster Reduction Center under the Ministry of Emergency Management.

Drought was widespread but not severe on the whole. The seriously affected provinces and municipalities were Jiangxi, Anhui, Fujian, Hubei, Hunan and Chongqing. The affected area of farmland was 8, 777, 800 ha and areas with no harvest reached 4, 179, 900 ha and direct economic losses of 37. 68 billion Yuan. A total of 6. 923 million urban and rural residents and 3. 681 million man-feed big animals and livestock suffered from temporary drinking water shortage.

Flood or drought affected and damaged areas

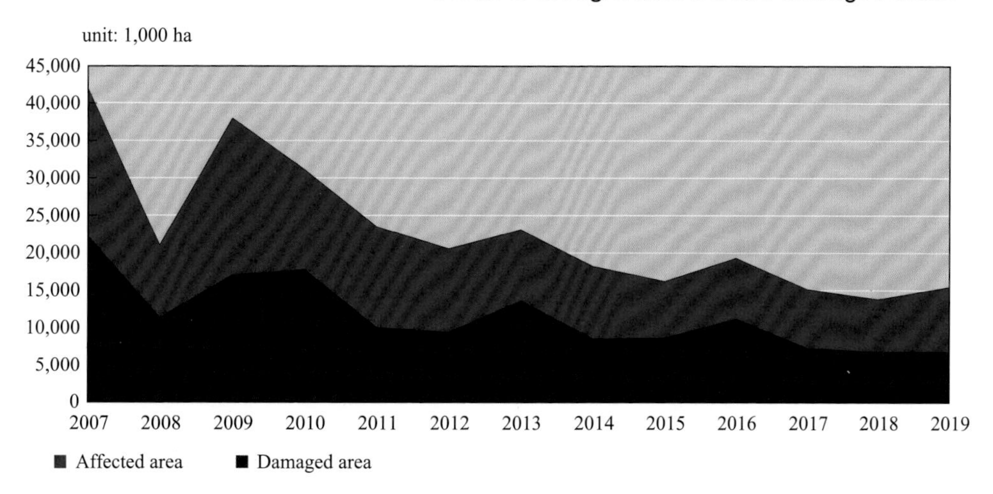

In 2019, the central government allocated a total of 2. 91 billion Yuan for water-related disaster relief, including 1. 94 billion Yuan for floods and 970 million Yuan for droughts. Up to 23, 553, 000 ha of cultivated land were irrigated against droughts, reducing loss of grain harvest by 24. 91 billion kg. Drinking water was provided to 6. 698 million rural and urban population and 2. 897 million big animals and livestock in order to alleviate temporary water shortage.

VI. Water Management and Reform

Water conservation management. The creation of water-saving society had been widely extended at the county level. There were 266 counties (districts) in 19 provinces or autonomous regions or municipalities such as Beijing Passed the final evaluation. The initiative of signing contracts for reducing water consumption in colleges and universities attracted about 130 million Yuan from the private sector. After signing of contracts with 40 universities, the average water saving rate was up to 20%. Within the water sector, a program was launched for government agencies to build themselves into water-efficient institutions. There were 43 affiliated units (including the departments of the Ministry of Water Resources) and provincial water departments met the criteria of water-saving type organizations, with a reduction of approximately 271, 000 m^3 of water consumption annually or 29% of the total.

River (lake) chief system. In 2019, all of the 31 provinces, autonomous regions and municipalities had CPC and government leaders serving as river (lake) chiefs. More than 300, 000 river (lake) chiefs were named at the four levels of province, city, county and township and more than 900, 000 river (lake) chiefs (including river inspectors and river guards) were named at the village level, achieving full coverage. River (lake) chief at provincial, city, county and township levels made a total of 7. 12 million river (lake) inspections and organized to implement the "one policy for one river/lake". A special action was initiated to address the issues such as misappropriation, illegal sand excavation, disposing of wastes and building structures without permission. A total of 137, 000 illegal activities were reported and corrected, with 35 million tons of garbage cleaned up in river courses, 39 million km^2 of illegal structures dismantled, 25, 000 km of banks freed from illegal use, 9, 800 km of enclosed embankments removed, 8, 600 km^2 of fishing nets removed, 10, 400 illegal sand mining sites and 9, 100 illegal sand vessels banned, which resulted in great improvement in rivers and lakes. In addition, a clean-up action was conducted to shorelines of the Yangtze River, identifying 2, 441 suspected illegal actions and 2, 230 of them were rectified. A special campaign was organized to reduce solid waste disposal in the Yangtze River Economic Belt, and clear up 1, 376 solid waste disposal sites within the administration of rivers and lakes. In collaboration with the Ministry of Transport and the Ministry of Public Security, joint actions, or the so called "clear the river" campaign, were taken to deal with sand mining along the trunk of the Yangtze River. Supervision and inspection were strengthened, as 6, 679 rivers and river sections and 1, 612 lakes in all cities with districts under their jurisdiction in 31 provinces, autonomous regions and municipalities were inspected without prior notice. Special investigations or inspections were made to oversee key river basins such as the Yangtze River, the Yellow River and the Grand Canal.

Most stringent water resources management. In 2019, the Ministry of Water Resources (MWR), in collaboration with the National Development and Reform Commission (NDRC) and other ministries, completed performance evaluation for the implementation of the most stringent water resources management system in 31 provinces, autonomous regions and municipalities in 2019. The 8 provinces, autonomous regions and municipalities of Jiangsu were rated as excellent. Water allocation plans of 5 cross-provincial river basins, including Wudinghe River, were approved. Water allocation was completed for river basins across cities or counties in 16 provinces, autonomous regions and municipalities, including Hunan. In the water diversion period of 2018 – 2019, Phase I of the Eastern Route of South-to-North Water Diversion diverted 0.844 billion m^3 of water to Shandong; Phase I of the Middle Route of South-to-North Water Diversion diverted 7.132 billion m^3 of water to Beijing, Tianjin, Hebei and Henan. As a result, water security in the receiving areas was greatly enhanced. Integrated water allocation in major cross-provincial river basins were highlighted and realized for the Hanjiang, the Jialing, the Wujiang, and 15 other cross-provincial rivers. The mainstream of the Yellow River had kept continuous flow for 20 consecutive years without dry up in the downstream. The East Dongjuyan Lake in the lower reaches of the Heihe River had been prevented from drying up for 15 consecutive years. The pilots for water ecological civilization saw smooth progress in 105 cities, of which 99 cities passed check and acceptance. Pilot projects were initiated in over-exploited areas of North China for groundwater recharge. A total of 949 million m^3 of water was released from the Middle Route of South-to-North Water Diversion and reservoirs in Hebei to three rivers in the Province helped reduce groundwater withdrawal. In 2019, China Water Exchange completed 237 entitlement trading with an amount of 115 million m^3.

Operation and management. In 2019, the central government allocated 1.45 billion Yuan as subsidies for small reservoir repair and maintenance, and local governments allocated 890 million Yuan as counterpart funding. The number of management staff trained for operation and maintenance of small reservoirs was up to 93,000. By the end of 2019, the number of approved national water scenic spots reached 878, including 373 reservoirs, 195 natural rivers and lakes, 195 lake or riverine cities, 47 wetlands, 31 irrigation districts and 37 soil conservation areas.

Water pricing reform. A survey was made to charges of water supplied by waterworks. Major issues related to water pricing were studied. National Development and Reform Commission improve water pricing and management system in collaboration with MWR. The revision and issue of Catalogue of Pricing by the Central Government allows for waterworks affiliated to the Ministry of Water Resources, which provided water at relatively low tariff that unchanged for a long period of time, to be listed in the plan of cost auditing. By the end of 2019, the reform of agricultural water pricing had been taken place in an area of 290 million mu of cultivated land, up by 130 million mu over the previous year.

Water resources planning and early-stage work. In 2019, there were 14 water resources plans approved by central government authorities (including printed and issued review comments). During the drafting of Water Security Plan of 14th Five-Year Plan period, participatory approach and online consultation had been adopted to request opinions from the general public. With approval of the State Council, the Action Plan for Comprehensive Control of Groundwater Overexploitation in North China was jointly issued by the Ministry of Finance (MOF), NDRC, and the Ministry of Agriculture and Rural Affairs (MARA). Substantial progress had been made in water resources planning under the National Ground Strategy, aiming at serving for coordinated development of the Beijing-Tianjin-Hebei region, development of the Yangtze River Economic Belt, integration of the Yangtze River Delta, construction of the Guangdong-Hong Kong- Macao Greater Bay Area, ecological protection and high-quality development of the Yellow River Basin, and further reform and opening up of Hainan Province. Comprehensive planning for major river basins and tributaries was accelerated with approval given to the plans for Taohe, Yiluo, Kuye, Xiangjiang and Zishui rivers. In 2019, a total of 14 projects subject to NDRC approval were finalized, including 13 feasibility study reports and 1 project plan, with a total investment of 41. 147 billion Yuan. Up to 12 preliminary designs were approved by MWR, with a total investment of 95. 779 billion Yuan.

Soil and water conservation. In 2019, MWR formulated and issued 13 rules and regulations including *the Opinions on Further Promoting the Reform to Delegate Power, Streamline Administration and Improve Government Services for Strengthening Soil and Water Conservation Monitoring and Supervision.* The approved soil and water conservation plans of construction projects was totaled 55, 900, covering a scope of 20, 800 km^2 for protecting and controlling soil and water loss. Water and soil conservation facilities of 13, 600 projects completed self-check and acceptance for filing purpose. Innovative measures were adopted to monitor man-made soil and water loss of production and construction projects through remote sensing, which covered an areas of 5. 92 million km^2. According to on-site review based on interpretation of remote sensing satellite, 53, 000 projects commenced without approval or had damped wastes before approval were investigated and punished.

Rural hydropower management. By the end of 2019, the title of green small hydropower station was awarded to 338 projects in 21 provinces. Standards for safe production and operation had been applied to hydropower stations in rural areas. The completed hydropower stations that complied with relevant standards accumulated to 2, 789 in the whole country, including 82 level one, 1, 168 level two and 1, 539 level three hydropower stations. Rehabilitation was completed for 1, 168

rivers, 1, 869 ecological restoration projects and 1, 997 expansion projects for efficiency increase. Over 2, 500 km of river courses were recovered from low water level or drying up.

Resettlement of water projects. In 2019, the resettled population was 217, 000, including 192, 000 rural residents and 25, 000 urban residents. Employment was arranged for 173, 000 resettled people. According to the data approved by the central government, the relocated people in rural areas due to the construction of large and medium-sized reservoirs in 2018, who should receive later stage compensation of governments, were totaled 113, 000.

Safety Supervision. In 2019, supervision, inspection and evaluation activities fall into three major categories: the implementation of river (lake) chief system nationwide, the most stringent water resources management system, and strengthening supervision of the water sector. A total of 2, 091 inspection teams with 8, 412 members were dispatched for 31, 049 projects, identifying 49, 629 problems. Up to 1, 741 inspections were made in line with the principles of "going straight to the sites without notice, personnel contact, reporting and accompany", accounting for 83. 3% of the total. A total of 45 accountability notices at or above the level of "verbal warning" were issued. There were 834 times of "verbal warning" or punishments above this level were taken against 759 enterprises who had failed to comply with the relevant stipulations. There were 8 production accidents with 11 people dead. MWR organized inspections on safe production of water project construction, concentrated rectification for safe production in the water sector, production safety patrols and major quality and safety inspection tours, and identified 47, 236 potential hazards with 97. 8% properly handled. Attempts were made to combine safety supervision with information technologies that enabled 48, 000 organizations fit into one safety monitoring platform. There were 103 organizations pass the safety inspection in accordance with standards of water project construction. There were 64 directly affiliated organizations met safety standards. *The Guidelines for Hazard Identification and Risk Evaluation in Operation of Water and Hydropower Projects* (*Reservoirs and Water Gates*) (*Trial*) was came into effect. Over the past year, there were 7 rounds of discipline inspections, with 100 teams and 815 members dispatched to 252 projects. More than 6, 000 violating activities were identified with issue of 99 notifications for rectification. Self-inspection of provincial water departments had been conducted for 170 times. A total of 3, 000 people divided into 500 inspection groups were sent out for investigation of nearly 2, 000 projects, and more than 16, 000 violations were found.

Legislation and administrative law enforcement. In 2019, 4 regulations and 2 normative documents were revised and 1 abolished. In 2019, the number of water-related cases investigated totaled 25, 115 and 22, 927 cases or 91. 3% were resolved. MWR handled and concluded 23 reconsideration and 7 responses to administrative cases.

Administrative permits. In 2019, MWR handled 1, 667 applications for water-related administrative approvals or permits with 1, 328 completed, including 19 project plan approvals, 12 preliminary design reports of water construction projects, 403 water abstraction licenses, 33 evaluation reports of flood impact by non-flood control project, 282 plans of construction projects within the jurisdiction of river courses, 90 approvals of soil and water conservation plan of production and construction projects, 27 approvals for establishment and adjustment of national basic hydrological stations, 27 approvals of hydrological monitoring projects for evaluating impact of construction at upper and lower of the national basic hydrological stations, 95 qualification approvals (including new application, extension, adding of new items or promotion) for construction supervisors of water resources projects; and 360 Grade-A qualification identifications (including new application, extension, adding of new items or promotion) for quality supervisors of water-related projects.

Water science and technology. In 2019, a total of 101 million Yuan was allocated to water-related science and technology projects, including 62. 73 million Yuan for three special-subject and water-related scientific research projects listed in the National Key Research and Development Plan-Effective Development and Utilization of Water Resources and Monitoring and Early Warning and Prevention of Major Natural Disasters. A total of 38. 1283 million Yuan was allocated to 57 demonstration projects for water technologies. Yangtze River Research Fund, founded jointly by MWR, National Natural Science Foundation of China and China Three Gorges Corporation, raised 250 million Yuan. A total of 21 research projects on major water-related technologies were launched. One water technological achievements won Special Prize and two won Second Prize of National Sci-Tech Advance Award. By the end of 2019, the number of national level or ministerial level labs were 12, and that of technical research centers were 15. There were also 6 ministerial field observation stations. A total of 139. 854 million Yuan was allocated from the central government as operation expenses for basic scientific studies of public research institutes. There were 33 water-related technical standards made public and 141 standards under drafting. By the end of the year, the number of effective water-related technical standards reached 854.

International cooperation. In 2019, a total of 6 water-related international cooperation agreements were signed. Up to 13 multilateral high-level roundtables or technical exchange seminars were held in China. 5 foreign-funded projects financed by the Asian Development Bank and Global Environment Fund were in smooth progress. Bilateral cooperation projects between China and Switzerland, Denmark, France, Finland and Netherlands, as well as foreign aid science and technology cooperation projects all made steady progress.

VII. Current Status of the Water Sector

Water-related institutions. By the end of 2019, there were 23, 554 legal entities, with 938, 000 employees and separate accounts, engaged in water-related activities within the administrative jurisdiction at county level or above. Among which, the number of governmental organizations were 2, 728 with 124, 000 employees, down by 1. 6% over the previous year; public organizations were 16, 145 with 500, 000 employees, down by 15. 0%; enterprises were 3, 758 with 306, 000 employees, down by 11. 0%; societies and other institutions were 923 with 7, 000 employees, down by 56. 3%. There were 27 general construction contractors awarded the highest qualification for water resources and hydropower project construction and 264 general construction contractors awarded Grade-I qualification.

Employees and salaries. Employees of the water sector totaled 854, 000, down 5. 4% from the previous year. Of which, in-service staff members amounted to 827, 000, down 5. 9%, including 66, 000 working in agencies directly under the Ministry of Water Resources, up 0. 6% over the previous year; and 760, 000 working in local agencies, down 6. 5%. The total salary of in-service staff members nationwide was 78. 76 billion Yuan, and the annual average salary per person was 95, 000 Yuan.

Employees and Salaries

	2009	2010	2011	2012	2013	2014	2015	2016	2017	2018	2019
Number of in service staff /10^4 persons	103.7	106.6	102.5	103.4	100.5	97.1	94.7	92.5	90.4	87.9	82.7
Of which: staff of MWR and agencies under MWR/10^4 persons	7.2	7.4	7.5	7.4	7.0	6.7	6.6	6.4	6.4	6.6	6.6
Local agencies/10^4 persons	96.5	96.3	95.0	96.0	93.5	90.4	88.1	86.1	84.0	81.3	76.0
Salary of in-service staff/10^8 Yuan	264.7	297.9	351.4	389.1	415.3	451.4	529.4	640.5	739.1	802.7	787.6
Average salary /(Yuan/person)	25,633	28,816	34,283	37,692	41,453	46,569	55,870	69,377	83,534	91,307	95,385

Main Index of National Water Resources Development (2014－2019)

Indicators	unit	2014	2015	2016	2017	2018	2019
1. Irrigated area	10^3 ha	70,652	72,061	73,177	73,946	74,542	75,034
2. Farmland irrigated area	10^3 ha	64,540	65,873	67,141	67,816	68,272	68,679
Newly-increased in 2019	10^3 ha	1,648	1,798	1,561	1,070	828	780
3. Water-saving irrigated area	10^3 ha	29,019	31,060	32,847	34,319	36,135	37,059
Highly-efficient water-saving irrigated area	10^3 ha	16,114	17,923	19,405	20,551	21,903	22,640
4. Irrigation districts over 10,000 mu	Unit	7,709	7,773	7,806	7,839	7,881	7,884
Irrigation districts over 300,000 mu	Unit	456	456	458	458	461	460
Farmland irrigated areas in irrigation districts over 10,000 mu	10^3 ha	30,256	32,302	33,045	33,262	33,324	33,501

Continued

Indicators	unit	2014	2015	2016	2017	2018	2019
Farmland irrigated areas in irrigation districts over 300, 000 mu	10^3 ha	11, 251	17, 686	17, 765	17, 840	17, 799	17, 994
5. Rural population accessible to safe drinking water	%		76	79	80	81	82
Centralized water supply system	%		82	84	85	86	87
6. Flooded or waterlogging area under control	10^3 ha	22, 369	22, 713	23, 067	23, 824	24, 262	24, 530
7. Controlled or improved eroded area	10^4 km^2	111. 6	115. 5	120. 4	125. 8	131. 5	137. 3
Newly-increased	10^4 km^2	5. 5	5. 4	5. 6	5. 9	6. 4	6. 7
8. Reservoirs	Unit	97, 735	97, 988	98, 460	98, 795	98, 822	98, 112
Large-sized	Unit	697	707	720	732	736	744
Medium-sized	Unit	3, 799	3, 844	3, 890	3, 934	3, 954	3, 978
Total storage capacity	10^8 m^3	8, 394	8, 581	8, 967	9, 035	8, 953	8, 983
Large-sized	10^8 m^3	6, 617	6, 812	7, 166	7, 210	7, 117	7, 150
Medium-sized	10^8 m^3	1, 075	1, 068	1, 096	1, 117	1, 126	1, 127
9. Total water supply capacity of water projects in a year	10^8 m^3	6, 095	6, 103	6, 040	6, 043	6, 039	6, 021
10. Length of dikes and embankments	10^4 km	28. 4	29. 1	29. 9	30. 6	31. 2	32. 0
Cultivated land under protection	10^3 ha	42, 794	40, 844	41, 087	40, 946	41, 351	41, 903
Population under protection	10^3 people	58, 584	58, 608	59, 468	60, 557	62, 785	64, 168
11. Total water gates	Unit	98, 686	103, 964	105, 283	103, 878	104, 403	103, 575
Large-sized	Unit	875	888	892	892	897	892
12. Total installed capacity by the end of the year	10^4 kW	30, 183	31, 937	33, 153	34, 168	35, 226	35, 564
Yearly power generation	10^8 kW · h	10, 661	11, 143	11, 815	11, 967	12, 329	12, 991
13. Installed capacity of rural hydropower by the end of the year	10^4 kW	7, 322	7, 583	7, 791	7, 927	8, 044	8, 144

Continued

Indicators	unit	2014	2015	2016	2017	2018	2019
Yearly power generation	10^8 kW · h	2, 281	2, 351	2, 682	2, 477	2, 346	2, 533
14. Completed investment of water projects	10^8 Yuan	4, 083. 1	5, 452. 2	6, 099. 6	7, 132. 4	6, 602. 6	6, 711. 7
Divided by different sources							
(1) Central government investment	10^8 Yuan	1, 648. 5	2, 231. 2	1, 679. 2	1, 757. 1	1, 752. 7	1, 751. 1
(2) Local government investment	10^8 Yuan	1, 862. 5	2, 554. 6	2, 898. 2	3, 578. 2	3, 259. 6	3, 487. 9
(3) Domestic loan	10^8 Yuan	299. 6	338. 6	879. 6	925. 8	752. 5	636. 3
(4) Foreign funds	10^8 Yuan	4. 3	7. 6	7. 0	8. 0	4. 9	5. 7
(5) Enterprises and private investment	10^8 Yuan	89. 9	187. 9	424. 7	600. 8	565. 1	588. 0
(6) Bonds	10^8 Yuan	1. 7	0. 4	3. 8	26. 5	41. 6	10. 0
(7) Other sources	10^8 Yuan	176. 5	131. 7	207. 1	235. 9	226. 3	232. 8
Divided by different purposes:							
(1) Flood control	10^8 Yuan	1, 522. 6	1, 930. 3	2, 077. 0	2, 438. 8	2, 175. 4	2, 289. 8
(2) Water resources	10^8 Yuan	1, 852. 2	2, 708. 3	2, 585. 2	2, 704. 9	2, 550. 0	2, 448. 3
(3) Soil and water conservation and ecological recovery	10^8 Yuan	141. 3	192. 9	403. 7	682. 6	741. 4	913. 4
(4) Hydropower	10^8 Yuan	216. 9	152. 1	166. 6	145. 8	121. 0	106. 7
(5) Capacity building	10^8 Yuan	40. 9	29. 2	56. 9	31. 5	47. 0	63. 4
(6) Early-stage work	10^8 Yuan	65. 1	101. 9	174. 0	181. 2	132. 0	132. 7
(7) Others	10^8 Yuan	244. 2	337. 5	636. 2	947. 5	835. 8	757. 4

Notes:

1. The data in this bulletin do not include those of Hong Kong, Macao and Taiwan.
2. Key indicators for water development and statistical data in 2012 and in 2013 is also integrated with the data of first national census for water.
3. Statistics of rural hydropower refer to the hydropower stations with an installed capacity of 50, 000 kW or lower than 50, 000 kW.